The Dodo /
JUV 598.65 San

Sanford, William R.
HAMBURG TOWNSHIP LIBRARY

29089

D1713754

DISCARD

# GONE FOREVER

THE DODO

by William R. Sanford
and Carl R. Green

CRESTWOOD HOUSE
New York

LIBRARY OF CONGRESS CATALOGING IN PUBLICATION DATA

Sanford, William R.
The dodo / by William R. Sanford and Carl R. Green
p. cm. – (Gone forever)
Includes index.
SUMMARY: Discusses the flightless bird that lived on the island of Mauritius and why it became extinct in the 1600s. Also describes the efforts of conservationists to protect endangered birds.
1. Dodo – Juvenile literature. [1. Dodo. 2. Extinct birds. 3. Wildlife conservation.] I. Green, Carl R. II. Title. III. Series.
QE872.C7S26   1989   598'.65–dc20   89-7867
ISBN 0-89686-455-3   CIP
                                                                            AC

Photo Credits

Culver Pictures, Inc.: 5, 11, 20, 38
DRK Photo: (Stephen J. Krasemann) 6, 7, 8, 13, 15, 16, 36; (M. P. Kahl) 9, 12; (Leonard Lee Rue III) 17; (Jeff Foott) 18, 43; (Don & Pat Valenti) 29; (Len Rue, Jr.) 41
The Academy of Natural Sciences of Philadelphia: (Steven Holt) 19
Photo Researchers: (Tom McHugh) 22
The Bettmann Archive: 27, 33

Cover Illustration by Kristi Schaeppi

Consultant: Professor Robert E. Sloan, Paleontologist
University of Minnesota

Copyright © 1989 by CRESTWOOD HOUSE, Macmillan Publishing Company

All rights reserved. No part of this book may be reproduced or transmitted in any form or by any means, electronic or mechanical, including photocopying, recording, or by any information storage and retrieval system, without permission in writing from the Publisher.

Macmillan Publishing Company
866 Third Avenue
New York, NY 10022
Collier Macmillan Canada, Inc.

Produced by Carnival Enterprises

Printed in the United States of America

First Edition

10   9   8   7   6   5   4   3   2   1

Contents

A Short History of Birds 5
Flightless Birds 10
What Kind of Bird Was the Dodo? 14
A Close Look at the Dodo 21
The Life of the Dodo. 24
The Dodo's Island Habitat 26
Europeans Discover the Dodo 30
The Path to Extinction. 32
Did the Dodo Really Exist? 35
Aiding Endangered Species 39
Other Birds in Danger? 41
Some Final Lessons 43
For More Information 46
Glossary/Index. 47-48

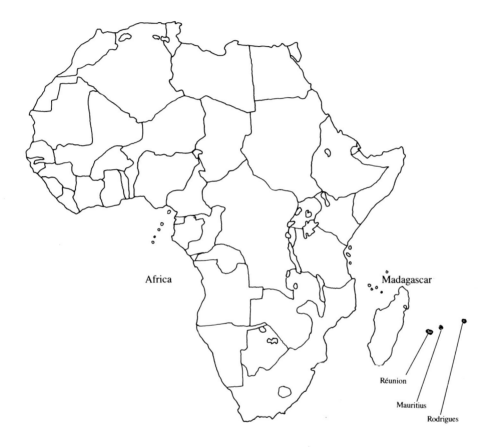

Before it became extinct, the Dodo was found on three small islands east of Madagascar near the southern tip of Africa.

Dodos were flightless birds that were not very attractive.

A Short History of Birds

At first glance, the Dodo looks positively silly. With its heavy body and wispy feathers, it's neither streamlined nor graceful. Other birds have given up the power of flight, but they have gained something in return. The ostrich became a powerful runner. The penguin gained the ability to swim and to dive for fish. *Ornithologists* (the scientists who study birds) don't laugh at the Dodo, however. They know it's a part of a story that's millions of years old.

Pterodactyls, the first flying animals, were cousins of the dinosaurs.

The first animals to take to the air were featherless, cold-blooded *reptiles*. Flight was a useful *adaptation*. It helped animals to escape from *predators* and to get food that other animals can't reach. The first flying animals were cousins of the dinosaurs called *Pterodactyls* (pronounced tero-DAK-tils). At one time, the skies were filled with the flapping of their wide, leathery wings.

Then, about 140 million years ago, a bird called the *Archaeopteryx* (pronounced ar-key-OP-terix) appeared. It still had a dinosaur's teeth and tail, but feathers protected it from the cold and gave lift to its wings. Scientists are still arguing over the flying ability of the *Archaeopteryx*. A few say its body was too heavy to do much more than glide from tree to tree like a flying squirrel. Most scientists refuse to accept this. They say that *Archaeopteryx*'s wings and feathers were as well developed as those of modern birds.

Animals adapt to meet the needs of their environment. The ostrich developed long legs in order to run faster than its predators.

The Great Egret is part of the carinate group, which are birds with excellent flight ability.

Before true birds emerged, several changes took place. First, early birds reduced their weight by developing strong, light bones. Beaks took the place of teeth, and saved a few more ounces. Next, birds developed strong wing muscles. As the muscles increased in size, the breastbone became a keel to anchor them. These changes developed over many generations.

In the centuries that followed, birds spread across the face of the earth. As their flying skills improved, they flew over mountains and across oceans. One major group of birds, the *carinates*, evolved into fine flying machines. Another, smaller group spent less and less time in the air. These were the *ratites*, the flightless birds. Some became

Antarctic penguins are ratites, which means they are birds that cannot fly.

seabirds and lived by diving for fish. Others found *habitats* on land after the dinosaurs died out. In remote places where four-legged predators were scarce, two-legged, meat-eating birds evolved.

One of the earliest ratites was *Diatryma*. The *fossils* of this fearsome bird are found in both North America and in Europe. *Diatryman* stood six feet tall, and its feet were tipped with sharp claws. As its size increased, this huge bird lost the power of flight. Its head was as big as a small horse's head, and it had a sharp, powerful beak. The short-legged *Diatryma* wasn't very fast, but it was agile. With rapid darts of its head, it speared the small animals that were part of its diet.

Diatryma was only one of many birds that gave up the skies. Where conditions were right, other flightless birds developed.

Flightless Birds

Most species of birds live in many different habitats. San Juan Capistrano's Swallows, for example, fly to South America each winter and return to California in the spring. The flightless Dodo and its cousins, by contrast, lived only on three small islands in the Indian Ocean.

Ornithologists don't find the Dodo's story too unusual. When conditions are right, birds seem to give up flying rather quickly. The first condition is a place that's nearly free of predators. If the bird doesn't have to fly from danger,

When birds such as the Dodo have few predators and a large food supply, they quickly lose their ability to fly.

flight becomes less important. In addition, the habitat must provide an abundant food supply. Birds that can find food close at hand aren't likely to do much flying.

As the birds spend less time in the air, nature takes a hand. The larger birds may be weak flyers, but they domi-

11

nate the battle for mates. The smaller birds cannot mate, and the species grows larger with each generation. Wings that aren't being used grow smaller. The breastbone loses its keel. For a while, the birds can still flutter for short distances. Then the day comes when they've lost the ability to fly.

Some birds make special adaptations to their new habitat. The penguins of the Antarctic are descended from birds that lived by diving for fish. Wings adapted to swimming and diving can't provide the lift needed for flight. Ostriches, on the other hand, lived on open plains where running speed was useful. They developed long legs and a

The wings of Antarctic penguins have become better adapted to swimming and diving than to flying.

Instead of developing their wings, ostriches have developed stronger legs to run on the open plains.

top speed of about 40 miles per hour. Of all ratites, the Kiwi of New Zealand may be the strangest. Instead of flying, these birds sought safety by burrowing into the ground. Now, thousands of years later, the Kiwi's two-inch wings are hidden by its feathers.

The Dodo's story follows a similar script. About 35 million years ago, a bird related to today's pigeons lived in Africa. Flocks of these birds flew eastward to three small islands now known as the Mascarenes. The trip of over a thousand miles took the birds to an island paradise. Food was plentiful and there were no predators. The birds evolved into three related species of ratites.

The Grey Dodo *(Raphus cucullatus)* is the best known of the birds of the Mascarenes. It lived on the island of Mauritius. The White Dodo (*Raphus solitarius*) developed on Réunion. Little is known about this bird. Except for its color, most ornithologists believe it was very much like the Grey Dodo. Perhaps it was only an *albino* form of the Mauritius Dodo. As an albino, it would have lacked the pigments that add color to a bird's feathers.

The third Dodolike bird lived on tiny Rodrigues Island. Because early settlers usually saw only single birds, they called them Solitaries. This fact is reflected in the bird's scientific name, *Pezophaps solitarius*. The Solitary was about the same size as the Dodo, but it had a longer neck and a shorter bill. It could run faster than the Dodo, but that didn't save it. Sailors who hunted the Solitary said it tasted better than its two cousins.

Naturalists were slow in reaching the Mascarenes. By the time they got there, all three species were *extinct*. With little firsthand information to go by, no one knew how to classify the Dodo and its relatives.

What Kind of Bird Was the Dodo?

The problem of classifying the Dodo started in the 1700s. At that time, there was no system for naming plants and animals. The chaos ended when Carolus Linnaeus, a Swedish naturalist, designed a system of Latin names for classifying all living things.

When a new bird is discovered, the person who names it must follow Linnaeus's rules. The bird's first name places it in its proper genus (the larger group to which the bird belongs). The second name identifies the species. Whenever possible, the names are chosen to describe the bird's looks or behavior. Under this system, all birds fit into one of 158 families.

The Dodo was first given the Latin name of *Raphus cucullatus*. The name means "cuckoolike bird with seams." It didn't seem to bother anyone that the name was

This museum's display includes bones of the Dodo that were discovered after the birds became extinct.

Later, the Dodo was put in the same family as the vulture.

totally off the mark. Dodos have nothing in common with cuckoos and they certainly don't have seams! Linnaeus tried to solve the problem by calling the bird *Didus ineptus,* or "clumsy Dodo." That was a better name, but the earlier name stuck.

Scientists also argued about the Dodo's relationship with other birds. Linnaeus put the Dodo in the ostrich family, despite its short legs. In 1835, a French naturalist moved the Dodo into the vulture family. He reasoned that Dodos and vultures both have hooked beaks. The fact that Dodos weren't fast-flying, meat-eating birds didn't bother him.

At first, scientists put the Dodo in the same family as the ostrich.

Some ornithologists put the Dodo in the same family as the Whooping Crane.

Were Dodos a type of vulture? Other naturalists went back to the early reports from the Mascarenes. The explorers said clearly that Dodos and Solitaries were plant eaters, not *scavengers*. With that, the Dodo was removed from the vulture family. Some ornithologists tried to put it in the same family as the penguin. Others said that it was related to the graceful, long-legged crane and ibis. None of these scientists had ever seen a Dodo.

In the 1840s an English ornithologist named Hugh Strickland began a more careful study. He found there were no live Dodos left, or even any skeletons. The best that Strickland could do was examine some paintings of the Dodo. Afterward, he wrote a book about what he had learned. The Dodo, Strickland concluded, was a *Columbiforme*. It was a relative of doves and pigeons. The Dodo's ancestors had flown to the Mascarenes millions of years ago.

Strickland's book started a new battle among ornithologists. Some agreed with him. Others laughed at the idea that the large Dodo, with its oversized beak, could be related to the smaller doves and pigeons. Strickland died shortly before evidence was found that proved his theory. On the

In the 1840s, ornithologists correctly classified the Dodo as a relative of doves and pigeons, like the Mourning Dove pictured here.

The Tooth-Billed Pigeon, found on the island of Samoa

Pacific island of Samoa, a large bird with a hooked beak (the Tooth-Billed Pigeon) was proved to be a member of the Columbiforme. Now, naturalists could agree that the Dodo and the pigeon had a common ancestor.

The final proof came in the 1860s. After a long search, Dodo skeletons were found on Mauritius. Experts studied the bones and concluded that the Dodo was a type of giant pigeon. After landing on Mauritius, an ancient, strong-winged pigeon had developed into the clumsy, flightless Dodo.

A Close Look at the Dodo

When the traveler Peter Mundy visited Mauritius in 1634, he saw his first Dodo. He wrote: "Dodoes, a strange kinde of a fowle, twice as bigg as a Goose, that can neither flye nor swymm...there being none such in any part of the world yett to be found."

Mundy's description isn't far from that of Lewis Carroll in *Alice in Wonderland*. Carroll's Dodo is absurdly fat. He carries a walking stick and takes forever to answer a question.

Because the Dodo became extinct before naturalists could study it, they have had to recreate the Dodo from skeletons and drawings. The results show that the Mauritius Dodo was larger than a Thanksgiving turkey, and just as fat. Male Dodos (*cocks*) stood about three feet tall. The females (*hens*) were slightly smaller.

As big as the Dodo was, the head still looked too large for the body. The eyes were round and the long, hooked beak was strong and sharp. Sailors who landed on Mauritius reported that those wicked-looking beaks could inflict painful bites. Some reports say only cocks had the heavily hooked beaks. The neck was one of the major differences between the Dodo and the Solitary. The Dodo had a very short neck. The Solitary's neck was long and graceful.

The best guesses place the Dodo's weight at around 50 pounds. Whatever its weight, the Dodo's fat belly often

bumped the ground when it ran. In some drawings, the bird looks fatter than in others. Naturalists say this may have been a part of the bird's breeding cycle. According to this theory, the Dodo was at its heaviest when it was nesting. Other scientists disagree, suggesting that the pictures show the Dodo at different stages of its yearly *molt*. Any bird, they point out, looks thinner after shedding most of its feathers.

There was nothing sleek or glossy about the Dodo. The Dodo's short feathers looked like the downy *plumage* of a *chick*. The skin around the eyes and beak was bare, and the tail contained only three to six fluffy, off-white feathers. White and gray were the Dodo's dominant colors. Its upper body was gray, its thighs were dark gray, and its belly was white. The beak was green or black, with a yellowish tip. The long feathers that tipped the Dodo's useless wings were yellow-gray. The upper wing feathers were white with black tips.

Visitors to Mauritius reported that Dodos used their stubby wings as weapons when they fought. The wings also made a rumbling noise when the birds flapped them against their bodies. Without the power of flight, the Dodo depended on its four-toed yellow feet to get around. With three toes pointing forward and a fourth turned back, it left tracks similar to those of a pigeon. Despite its size, the Dodo could move swiftly over rocky ground. Catching a Dodo was harder than one would think.

No one knows what the call of the Dodo sounded like. One guess is that the bird was named for the sound it made.

This wooden model of the Dodo is at the Field Museum in Chicago.

If that was true, the call must have been a honking "do-do-do." Whatever the sound, the woods of Mauritius once echoed with the noisy calls of thousands of Dodos.

The Life of the Dodo

When the Dodo's ancestors landed on Mauritius, they found an almost perfect habitat. Food was plentiful and enemies were nonexistent. It was the coming of the first Portuguese and Dutch sailors that doomed the Dodo. Faced with predators, both human and animal, the Dodo was almost helpless.

Until 1507, however, the Dodo was free to live its slow, peaceful life on Mauritius. Each year was much like the one before. In that tropical climate, winter was almost as warm as summer. Rainfall was plentiful, and the island produced a wealth of fruits and seeds. The Dodo feasted on its favorite foods, mated, and raised its chicks.

After the chicks left the nest, the Dodo went into its yearly molt. The old feathers fell out and new feathers grew to replace them. Growing feathers takes a lot of energy, and the birds lost weight during this time. The Dodo, never beautiful, must have looked like a partly plucked turkey during the molting season.

As the new feathers grew in, the cocks would *display* their plumage. When two cocks met, each bowed his head and shook his feathers as a form of greeting. During the mating season, displays became formal "dances." When a hen looked interested, the cock spread his small wings, jumped, and stomped his heavy feet. Then he pecked rapidly at the ground, picked up seeds, and tossed them in the air. Soft, crooning sounds came from his throat. If the hen wasn't interested, she wandered away to find another eager dancer.

At times, two cocks probably fought for the right to mate with a prize hen. On Rodrigues Island, François Leguat saw the male Solitaries battle for mates. The cocks whirled round and round, flapping their wings and making loud noises. Then they ran at each other, striking out with their wings. Each of the Solitary's wings contained a round mass of bone at the joint that made a good weapon. Leguat says that the bone was "as big as a musket ball."

Unlike the more social Dodos, Solitaries lived in a specific territory. Each mated pair defended an area that extended about 200 yards from the nest in all directions. If another bird came near, the Solitaries spun around in circles and flapped their wings. The flapping produced a loud, rattling noise that warned the intruder to stay away. Fights broke out if the strange bird came too close to the nest.

Once mated, Dodos found quiet spots in the forest to build their nests. The large nests were made from matted grass. The hen laid a single white egg, and both birds took turns sitting on it. No Dodo eggs exist today, but they were probably about the size of small grapefruit. One observer claimed to have seen a Dodo egg that was six inches long.

After the chicks hatched, they needed a long period of

care by the parent birds. Like other members of the pigeon family, Dodos probably fed the chicks with "pigeon's milk." This is a milklike fluid that pigeons produce in their *crops*. The rich "milk" gave the Dodo chicks a good start. As the chicks grew, the adult birds added fruits, seeds, and tender plants to the young birds' diet.

By the time they were a year old, the young Dodos were as big as their parents. They were ready to take their places as adult birds in their island habitat.

The Dodo's Island Habitat

The Dodo and its two cousins lived only on three small islands in the Indian Ocean. The islands, known today as the Mascarenes, are located 500 miles east of the large island of Madagascar. Mauritius, which is three-fourths the size of Rhode Island, is a little smaller than Réunion.

In 1507, Portuguese explorers landed on Mauritius and named it for their shop, the *Cerne*. The Dutch arrived a few years later, misread the name, and thought they were landing on the Island of Swans. One look at the Dodos told them that these fat birds weren't even close to being swans! The Dutch renamed the island in honor of their ruler, Count Maurice of Nassau. It has been known as Mauritius ever since.

The island became a favorite stopping place for ships in

This engraving of the Dodo was based on a painting in the British Museum in London.

need of water and supplies. Dutch settlers arrived in 1638, but they abandoned the colony 70 years later. The French took over the islands in 1721, only to lose Mauritius to the British in 1810. Today, Mauritius is a democratic member of the British Commonwealth. Its million-plus inhabitants grow spices, sugar cane, pineapple, and bananas.

All of the Mascarenes were formed by volcanoes. Mauritius is circled by coral reefs and has a fine, protected bay. Rainfall ranges from 40 inches in the northwest to 175 inches on the eastern slopes. Heavy runoffs have cut deep ravines into the hillsides. The highlands have a cool, pleasant climate, but the lowlands are hot and muggy. The entire

island supports a heavy growth of tropical trees, flowers, and other plants. It's not surprising that the early explorers thought Mauritius was the Garden of Eden.

The Dodos and other birds fed on seeds, fruits, and green plants. With its strong beak, the Dodo cut soft fruits into chunks and swallowed them. With tougher fruits, it held the fruit with its feet so that it could tear off pieces with its beak.

Hard-shelled nuts were another matter. To digest these, the Dodo had to depend on its *gizzard* (a strongly muscled section of the stomach). When the gizzard contracts, it grinds hard bits of food against gravel and small stones. Dodos, in fact, stored stones as big as walnuts in their gizzards.

Like food, water was plentiful on Mauritius. When the Dodo went to drink, it displayed another pigeonlike behavior. Instead of lifting its head to swallow, it sucked up water through its beak. Young birds received all the water they needed in the food their parents gave them.

Although they couldn't swim, Dodos often bathed in ponds. They waded into the water, fluffed out their feathers, and enjoyed a peaceful moment in the sun. After bathing, the Dodo looked for a quiet place to *preen* itself. It cleaned and oiled its feathers by drawing them through its beak. Then the Dodo ducked its head and cleaned its head and beak with its claws. At night, the Dodos nestled together in the long grass.

How many years did a Dodo live? No one knows for sure. Naturalists guess that the big birds had a long life span.

The Dodo's behavior was similar to the Passenger Pigeon's, which is also extinct.

Another flightless bird, the ostrich, sometimes lives for 60 years. A few Dodos survived long sea voyages and lived for several years in Europe's much harsher climate.

Europeans Discover the Dodo

Marco Polo returned from his adventures in the Far East in 1295. His tales of the wealth of the Indies created a new interest in trading with the Orient. Polo's city-state of Venice already controlled the land routes across Asia, however. Other European nations had to look for new sea routes. That was the search that sent Christopher Columbus sailing westward in 1492. Instead of reaching the spice islands, Columbus discovered the Americas.

While Columbus was carrying Spain's flag to the New World, sailors from Portugal were trying a different route. Portuguese ships pushed farther and farther down the west coast of Africa. The captains believed that if they could round the tip of the continent, the way to India would be open. Bartholomeu Dias reached the Cape of Good Hope, at the southern tip of Africa, in 1488. But it was Vasco da Gama who sailed on to India in 1498. The trip was long and dangerous, but the rewards were great. More Portuguese ships set out for India.

In 1507, Captain Diego Fernandes Pereira discovered three small islands east of Madagascar. Six years later,

Captain Pedro Mascarenhas stopped there, too. The Mascarenes were named in his honor, even though Pereira was the first to land.

Like all voyages of the time, the ships were at sea for many months. The sailors quickly ran out of fresh food and were forced to eat salt beef and dry biscuits. The biscuits were often full of worms. Any islands that could supply fresh water and meat were welcome.

Both Pereira and Mascarenhas picked up fresh stores on Mauritius. Neither mentioned the Dodo, although the sailors probably killed and ate some of the big birds. For most Europeans of the 1500s, wild animals were either useful or dangerous. If they were useful, people ate them or put them to work. If they were dangerous, people killed them. The idea that a species should be protected would never have entered the minds of hungry sailors.

The settlers who came to the Mascarenes didn't think the Dodo was anything special. More likely, they probably laughed and said the bird was rather stupid. Some naturalists, in fact, don't believe the Dodo's name was taken from its call. They think the name comes from the Portuguese word *doudo,* meaning "simpleton." The French, who couldn't make up their minds, called it either a wild turkey or a hooded swan. The first Dutch to arrive on Mauritius named it the *walghvogel,* the "disgusting bird." Somehow, "Dodo" won out.

The ships that stopped in the Mascarenes sometimes carried live birds away with them. When the Dodo reached Holland, the Dutch artists thought it was wonderful. They painted a number of pictures of the Dodo, many of which weren't very accurate. Around 1637, a Dodo arrived in England and lived there for a number of years. After the bird died, it was stuffed and put on display in a museum. In 1683, the stuffed Dodo was moved to the Ashmolean Museum in Oxford. No one knew at the time that the last Dodo on Mauritius was already dead.

The Dodo remained at Oxford until 1755. That was when the director decided to get rid of the battered old bird. The Dodo was thrown into a pile of trash, ready for burning. Luckily, someone saved the head and one foot from the fire. Today, except for some skeletons that were dug up later, that's all there is. No one ever saw another complete Dodo, dead or alive.

The Path to Extinction

The world learned about the Dodo from the journals of a Dutch explorer, Admiral Jacob van Neck. Van Neck's fleet of eight ships dropped anchor in the bay at Mauritius in 1598. The admiral was a careful observer. He described the island's doves, turtles, ebony trees, and Dodos (he was the one who named them *walghvogels)*. When the expedition sailed, it carried two live Dodos with it.

Van Neck's report was printed in 1601. The book was widely read, and people took a great interest in the "disgusting birds" of Mauritius. Van Neck explained that the name referred to the taste of a cooked Dodo, not its looks. The longer the bird was cooked, the worse it tasted.

Dodos were not used to predators and did not know how to protect themselves from the sailors who landed on their islands.

Only the breast and belly were worth eating, he explained.

The next sailors to reach Mauritius didn't share van Neck's opinion. A second Dutch expedition landed in 1601 and feasted on Dodo meat. The crew killed two dozen birds the first day and almost as many the next. The sailors didn't have to waste gunpowder on the peaceful birds. One man with a club could kill a dozen Dodos in a few minutes. The birds simply didn't know how to protect themselves. When the ships sailed away, they carried half a ton of salted Dodo meat with them. The birds were so big that two of them provided a hearty meal for the entire crew.

Mauritius became a favorite stopover. Ships filled their

33

water barrels and picked up fresh fruit and meat before sailing on to India. After settlers arrived in 1638, they made Dodos part of their diet. They also killed extra birds and preserved the meat. When a ship arrived in the bay, the settlers traded Dodo meat for clothing, gunpowder, and other goods.

If the settlers had been the only predators, the Dodo might have survived. The settlers, however, also turned their dogs loose. The dogs chased the adult Dodos off their nests and stole their eggs. When they weren't eating eggs, the dogs killed defenseless chicks. Rats, pigs, and macaques (a type of monkey) soon joined the dogs in feasting on Dodo eggs. The rats came ashore from visiting ships. The pigs and macaques were brought ashore by settlers. Every egg that was eaten was a disaster for the species. The loss of their egg meant that the parent birds wouldn't hatch a chick that year.

The number of Dodos fell quickly. On his second visit to Mauritius, Peter Mundy realized Dodos were becoming scarce. He had seen a number of birds only four years earlier. In 1638, he wrote: "We now mett with None." The last person to see a living Dodo may have been Benjamin Harry, who saw a few birds in 1681. Twelve years later, a search of the island failed to turn up a single Dodo. Réunion's White Dodo disappeared about the same time. The Rodrigues Solitary was last seen in 1761.

The loss of the Dodo was a new experience for the people of the 1600s. This was the first recorded extinction of an entire species. Even worse, naturalists hadn't had time to study the Dodo. The lack of good information about this bird raised a new question: Did the Dodo exist? Some people said yes, but others said no.

Did the Dodo Really Exist?

By 1800 the Dodo had been gone more than a hundred years. Some naturalists were wondering if the flightless birds of the Mascarenes had ever existed. Were they *mythical* creatures like dragons and winged horses?

Before accepting the Dodo, scientists wanted proof. They found many pictures, but very few of them looked alike. According to the artists, every Dodo was different. The birds in the paintings varied in the lengths of their legs, the sizes of their beaks, and the shapes of their bodies. Almost all physical traces of the Dodo had been lost.

Next, the naturalists turned to written reports. They read the diaries of van Neck, Mundy, Leguat, and others. Could they trust these explorers? After all, Marco Polo had described the roc as if it really existed. In a similar way, Ponce de Leon had chased all over Florida after the mythical Fountain of Youth. It seemed clear to the doubting naturalists that the explorers had been wrong. Perhaps they had seen a large goose or an ostrich, and had called it the Dodo.

J. S. Duncan, an Oxford naturalist, wrote a paper on the Dodo in 1828. The professor looked at the evidence and charged that the skeptics were wrong. The Dodo was extinct, Duncan concluded, but it had existed at one time. When Duncan's paper reached Mauritius, it created a lot of interest. Three men formed a natural history club and set out to find the bones of the Dodo. Despite a careful search, no bones were found. The digging on Mauritius, however, led to the discovery of some unusual bones on Rodrigues. The bones were sent to Europe, where they were identified as those of the Solitary. That was a step forward. If the Solitary was real, maybe the Dodo was, too.

Why couldn't the searchers find any Dodo bones? George Clark, who was born on Mauritius, came up with a clever theory. Clark said that the Dodos' bones had all been washed away by heavy tropical rains. If that was true, Clark asked himself, where would the bones have ended up? The map showed the answer. Mauritius's three small rivers meet in a marshy delta. It was a perfect place to look for Dodo bones.

In 1863, Clark hired workers to dig in the delta. The men soon turned up a large number of bones. Clark and other naturalists were able to assemble complete skeletons. Museums from all over the world put the bones on display. The Smithsonian Institution in Washington, D.C., has one skeleton, and the American Museum of Natural History in New York City has another.

In the New York display, a realistic model of the Dodo

All that remains of the Dodo today are a head, one foot, and bones, which are on display in some museums.

stands beside the skeleton. The feet and head were copied from the last remains of the Oxford Dodo. When it came to adding feathers, the model makers knew that no Dodo feathers had been found. They solved the problem by covering the model with feathers from birds that have a similar plumage.

The skeletons and models ended the debate about the Dodo's existence. The displays also answered another old question. Was the Dodo as awkward and silly-looking as the explorers and painters of the 1600s said it was? One look was enough to end that debate, too. The poor old Dodo was a funny-looking bird.

The loss of the Dodo has shown us how fragile a species can be.

Aiding Endangered Species

Modern scientists can perform near miracles, but they can't create life. *Endangered species,* therefore, must be saved before they become extinct. Once a species is gone, no one can bring it back. The Dodo is gone forever.

The forces that drive plants and animals to extinction are easy to find. In 1900, for example, there were less than two billion people on earth. Today there are more than five billion. Wherever these people live, they need food, housing, clothing, and a place to work. Wildlife habitat gives way to cities and factories, farms and ranches. The Dodo disappeared because no one cared enough to save it. Will that happen to other species if people don't try to save them?

The idea of rescuing endangered species is a modern one. It wasn't until late in the 1800s that people began to support the idea of *conservation.* The Audubon Society, founded in 1886, was one of the earliest groups to pursue this goal. The society emphasized the study and protection of birds, but it was 200 years too late to save the Dodo.

When the conservation movement began, a number of birds were being pushed toward extinction. Some were hunted for their plumage, and others for their meat. The egret, for example, was once in danger because hatmakers wanted its feathers. Only a change in fashion saved this lovely bird. The Audubon Society urged the states to protect endangered birds. President Theodore Roosevelt, although he was a big-game hunter, helped the cause. Roosevelt supported conservation and expanded the national parks system.

Thanks to the conservationists, wildlife now has a better chance. The U.S. parks system's 28 million acres provide every type of wildlife habitat. Around the world, other countries have joined in the movement to save the animals. Good intentions haven't been enough, however. In Africa, *poachers* (illegal hunters) slip into wildlife parks and kill animals for food or profit. In many countries, farmers destroy wildlife habitats in order to plant crops.

People all over the world are joining hands to reverse this trend. The World Wildlife Fund (WWF), founded in 1961, has become a leader in the conservation movement. The WWF relies on the support of both private citizens and national governments. With their help, it is fighting to save the gorilla, the giant panda, and other endangered animals.

Much good work has been done, but the problem hasn't gone away. The U.S. Fish and Wildlife Service keeps a list of endangered species. The roll call of endangered birds includes the Bald Eagle, the Ivory-Billed Woodpecker, and the Ruby-Throated Hummingbird. The long list also includes the Utah Prairie Dog, the Southern Sea Otter, the American Crocodile, and the Yaqui Catfish.

No one wants these animals to become extinct, but good intentions aren't enough. More animals will go the

The American Bald Eagle, a symbol of the United States, is on the list of endangered species.

way of the Dodo if we do not listen to the lessons of the past.

Other Birds in Danger?

When something is totally lifeless, people describe it by saying, "It's dead as a Dodo!" That is a sad way to be

41

remembered, but the Dodo was the first extinction in recorded history. It has become a symbol for humanity's careless treatment of animals.

Since the 1600s, other birds have joined the Dodo in extinction. One of the most notable losses was that of the Passenger Pigeon. The Grey Dodo only lived on one small island, but these beautiful pigeons ranged across most of North America. A single flock of Passenger Pigeons sometimes numbered as many as 50 million birds!

After the loss of the Passenger Pigeon, conservationists worked harder to save other birds. The Whooping Crane, for example, has become a success story. These long-legged white birds probably numbered only a few thousand when the settlers arrived in the Americas. The cranes were easy targets, and their numbers fell even faster when farmers began draining the marshes where they nested.

Today the number of wild Whooping Cranes has risen to over 50, from a low of only 18 in 1938. The birds are still endangered, but the species seems to be on its way back.

Rescuing the California Condor has been even more difficult. These big birds, with their nine-foot wingspan, live only in the mountains of Southern California. In 1988, there were only 27 Condors left. Naturalists at the San Diego Wild Animal Park were overjoyed, therefore, when they hatched a 6.75-ounce chick. Molloko was the first Condor to be hatched from an egg laid by a captive Condor. If more Condors can be raised at the park, the birds can one day be returned to the wild.

Conservationists have prevented the Whooping Crane and other birds from becoming extinct.

A similar program has saved the Peregrine Falcon, but the job isn't over. The last Dusky Seaside Sparrow died in 1987. The Mauritius Kestrel and the Mauritius Pink Pigeon are on the endangered list. The lessons of the Dodo and the Passenger Pigeon must not be forgotten.

Some Final Lessons

Paul Ehrlich is a naturalist and a leader in the conservation movement. He compares the loss of a single species to the loss of a rivet from an airplane. If a plane loses one or two or three rivets, it still flies safely. What happens if enough rivets are lost? The plane falls apart and crashes.

Ehrlich believes that a thousand species of plants and animals come close to extinction each year. By the year 2000, he fears, the number may reach 10,000 a year. That's one species lost for every hour that passes! "The more species lost," Ehrlich says, "the greater our chances for meeting an ecological disaster that will cause Spaceship Earth to crash."

Luckily, the loss of the Dodo didn't cause the ship to crash. But it did offer us a number of lessons.

Lesson 1: A long history is no guarantee against extinction. In one form or another, the Dodo existed for many thousands of years. It adjusted to changes in the climate, and it adapted to a favorable habitat on Mauritius. All was well with the Dodo until humans arrived on the island. Change comes slowly in nature, and the Dodo didn't have time to develop new wings.

Lesson 2: No species can survive the misuse of human technology. Some people argue that the Dodo was a type of "living fossil," just waiting for extinction. That's an unfair judgment. Nature could never have prepared the Dodo for the arrival of the first humans. For all its benefits, technology carries with it the threat of extinction.

Lesson 3: The loss of one species can lead to the loss of others. The loss of the Dodo started a chain reaction on Mauritius that is still going on. One of the island's large trees, the *Calvaria,* is dying out. The trees still produce fruit with fertile seeds, but the seeds don't grow. When the last of the old trees dies, the species will be extinct. Naturalists believe that the *Calvaria*'s hard, thick-shelled seeds can't sprout unless they've been through a Dodo's gizzard! The grinding action of the gizzard didn't crush the seeds, but it did wear them down. Then, when the seeds became part of the Dodo's droppings, they were ready to sprout.

Will humanity learn these lessons? The Dodo is gone, but the Whooping Crane and the California Condor are still with us. Let's try not to lose any more rivets from this wonderful "Spaceship Earth."

For More Information

For more information about the Dodo, write to:

    The Academy of Natural Sciences of Philadelphia
    19th & The Parkway
    Philadelphia, PA 19103

Glossary/Index

Adaptation 6, 12 – the physical and behavioral changes that a species goes through in adjusting to a new habitat.

Albino 14 – any plant or animal having abnormally white coloring.

Carinates 8 – birds that have the ability to fly.

Chick 23, 24, 25, 26, 42 – a young male or female Dodo.

Cocks 21, 24, 25 – adult male Dodos.

Conservation 39, 40, 42, 43 – the protection of our natural resources.

Crops 26 – sacklike spaces in birds' throats where food is softened for digestion.

Display 24 – to show off, to reveal.

Endangered species 39, 40, 41, 42 – an animal that is in danger of becoming extinct.

Extinct 14, 15, 21, 28, 32, 34, 37, 39, 40, 42, 44, 45 – no longer living.

Fossils 10, 44 – the remains of dead animals, left either as bones or other body parts, or the imprints of these parts in a rock.

Gizzard 28, 45 – an enlarged section of a bird's stomach, where grit and small stones grind up hard bits of food.

Habitats 10, 11, 12, 24, 26, 39, 40 – places where animals make their homes.

Hens 21, 25 – adult female Dodos.

Molt 23, 24 – the time when a bird's feathers fall out and are replaced by new ones.

Mythical 35 – imaginary or not real.
Naturalists 14, 17, 18, 20, 21, 23, 28, 31, 34, 35, 37, 42, 43 – scientists who study plants and animals.
Ornithologists 5, 10, 14, 18, 19 – scientists who study birds.
Plumage 23, 24, 38, 39 – a bird's coat of feathers.
Poachers 40 – hunters who break the law by killing protected animals.
Predators 6, 7, 10, 11, 14, 24, 33, 34 – animals that live by killing other animals.
Preen 28 – to clean and comb. When it preens, a bird uses its beak to spread oil on its body from a gland at the base of its tail.
Ratites 8, 10, 13, 14 – birds that have lost the ability to fly.
Reptile 6 – a class of cold-blooded animals with backbones and lungs. Reptiles usually have skin covered with horny plates or scales.
Scavengers 18 – animals that live by feeding on the carcasses of animals killed by predators.